Origin and Development Of Calendars in the World

By

Dr. Ravi Prakash Arya

AMAZON BOOKS, USA

In association with

INDIAN FOUNDATION FOR VEDIC SCIENCE

H.O.1051, Sector-1, Rohtak, Haryana, India Ph. 01262-292580
Delhi Contact Ph. Nos. + 91 9313033917; + 91 9650183260
Emails: vedicscience@rediffmail.com
vedicscience@hotmail.com
www.vedicscience.net

First Edition
Kali era: 5015 (c. 2014)
Kalpa era: 1,97,29,49,115
Brahma era: 15,50,21,97,9,49,115

ISBN No. 81-87710-71-3

My Beloved Father

Late Ram Narain Arya

Vedic Meteorologist

Who dedicated his life to social service and propagation of Vedic teachings and philosophy .

Origin of calendars is a subject that may interest one and all. Throughout the world, different calendars have been in vogue from time to time. Now everybody is following the Gregorian calendar. Before reaching the present stage, we had to pass through many stages to bring out the calendar to this perfect state. Here we shall take into account all facts underlying the origin and development of calendars in the world.

Today December is considered as the last month of the year throughout the world and so 1st January is celebrated as the beginning of the new year. So far as the Vedic culture in Bharat is concerned, it doesn't celebrate January as the beginning of the year, rather as per Bharatīya calculations, *Varṣa pratipadā* (first day of the year) falls either on the day of vernal equinox or winter solstice or summer solstice. As such a curious mind may a ask a question why there is a difference regarding the commencement of new year as per the Indian view and the view of the rest of world.

Here the answer lies in observing scientific and non-scientific attitudes. This difference also reflects the advancement of science and scientific awareness during the Vedic age. The Vedic people had enjoyed science and scientific notions to the extent that various rites, rituals, social traditions were developed and organised scientifically in the light of the scientific laws/awareness persisted in that period. Interestingly, in the Vedic period, science was described and defined by the term *Dharma*. *Dharma* had different connotations in different scientific contexts. It stood for the scientific laws sustaining universal creation so far as physical and astrophysical sciences are concerned. In the context of metaphysical sciences, *Dharma* denoted the moral and spiritual laws sustaining the bio-life in this universe. To the Vedic seer, everything that was in accordance with *Dharma* was acceptable and everything standing against *Dharma* was not acceptable. That is why there became prevalent the concepts of *Dharma* and *Adharma*. *Dharma* was never taken in the sense of religion or religious sects as we do it today. The above cited Vedic notion of *Dharma* has been pointed out by Yāska, one of the leading ancient Vedic scholars, in one of his observations as follows:

sākṣāt kṛtdharmāṇaḥ ṛṣayo babhūvuḥ

(*Nirukta*, 1.20)

"That is there were born *Ṛṣis* who visualised universal laws of nature in the form of Vedas."

Following the same Vedic notion of *Dharma* and *Adharma*, Dayananda Sarasvati, a living embodiment of Vedic life and thought in the middle of the 50th century of *Kali* era (19th century of Christ) announced in the fifth principle of Arya Samaj : "All the acts should be done in accordance with *Dharma* i.e. after deliberation of what is true (scientific) and what is not true (unscientific). Thus the concept of *Dharma* was having a notion of science behind it, and the same was the very basis of Vedic life and thought. That is the reason why all sorts of Vedic traditions and rituals reflect a kind of scientific attitude of

Vedic people and conveys some scientific significance behind them. The same thing holds true with regard to the traditions inherited by modern Indian society from their Vedic ancestors.

Now coming to the very question of celebrating the new year. I have no hesitation to say that howsoever scientifically advanced and developed has become the modern society, it has not come out of its narrow confines of communalism, regionalism, racism and fundamentalism. It is using science in the scientific field and in its personal or social life it is using sectarian and social biases and taboos howsoever unscientific they may be. The old paradigm (i.e. so far as science is concerned, the Earth moves around the Sun and so far as Bible is concerned, Sun moves around the Earth) is still haunting us. So called modern society is still living with double standards. On one hand we claim to be scientifically advance, on the other hand, we want to be dwarfed by the narrow sentiments of communalism, regionalism, racism and fundamentalism. We have not yet started living science in our personal and social life which is governed by the above slogans. Until and unless we do so, we will continue to fight and terrorise each other in the name of various -isms.

If we want to get rid of all these problems, we shall have to go back to the Vedas, which was once the way of life on the whole Globe. Vedic society was the society which lived a life according to *Dharma* (science). It was not ridden with any sort of taboos governed by any sort of -ism. It tried to live what was *Dhārmika* or scientific and shunned what was considered to be opposite to *Dharma*, i.e. unscientific.

We people in the modern age are undoubtedly using science for our convenience and comforts but lack scientific awareness and are living superstitions, delusions, biases of all kind, hypothetical and preconceived notions ignoring science. On the other hand, Vedic people had a scientific awakening. They enjoyed science as well as lived science in their lives. The modern world owes a lot to the Vedas and Vedic tradition in spite of its highest technological advancement. That is why, what cannot be known with the help of modern science, can easily be unravelled with the help of extant Vedic traditions.

The concept of a new year is the glaring example of the above observation. In the early periods, Vedic calendar system was uniformly prevalent all over the Globe, there being only one Vedic culture prevalent on the Earth. Later on when the Vedic people on the other parts of Globe began to lose their contact with their mainstream in Bharat for want of accessibility from one region to another, there started developing variations and changes in the actual Vedic heritage inherited by people on the various parts of the Globe owing to their particular personal, regional or other reasons. Thus looking back to Vedas we would be able to ascertain the origin and development of calendar system in the world.

As indicated earlier, the calendar system in the world was first discovered by Vedic people during the Vedic age. Vedic people were also the first discoverer of the time calculation system in the world. They discovered both solar and lunar calendars for the calculations of days, months and years. They developed a luni-solar calendar based on the harmonisation of lunar and solar cycles. This Vedic calendar was known as luni-solar or synodic lunar calendar, being based on the synthesis of both lunar and solar cycles. There are some scholars who have a doubt that the Indians did not know the solar months. That is why they measured the time period of a month based upon lunar orbit of the earth. The moon takes one complete orbit of the earth in 27.3 days with respect to constellations which is called a sidereal month. On the other hand, the moon takes one complete orbit of the earth in 29.5 days with respect to the sun. This period is called a synodic month. The Vedas takes into account a sidereal month because the human menstrual cycle corresponds with the sidereal month of 27 days. In the Vedas, there are many mantras which talk about the mother delivering a child in the tenth month- दशमे मासि सूतवे। This period of 10 months refers clearly to the sidereal month. The child is delivered in 27x10=270 days which is equal to 9 solar months. Thus, we have seen that Vedas are more practical and scientific while formulating concepts of various units of time. Since the sidereal month has its application in human physiology. So, this was accepted as a unit of the month. A synodic or solar month has no practical applicability. There was one more reason to accept the moon at the base of calculations of days and month. Different waxing and waning lunar phases, as well as Full moon day (Pūrṇimā) and New moon day (Amavasyā), are the natural phenomena actually happening in the universe. So, it was very easy to conceive the concept of the day (lunar day) forming a unit of dark and bright halves of a month. Moreover, it was very easy to give nomenclature to lunar days as first, second etc. of bright half and dark half. So, this was the reason for ancient Indians to choose the moon at the base of calculations pertaining day and month. Just as a lunar day is the natural unit of time, which can be comprehended naturally without any artificial means, so is the solar day and night. But, the real problem comes while measuring solar month. The end of the start of lunar month can be naturally comprehended with the occurring of Full moon day (Pūrṇmāsī) and New Moon day (Amāvasyā), but it is difficult to take stock of an end and start of a solar month. Before the final introduction of Julian reforms in the western calendar till 46 BC, it was not possible for the Europeans to calculate the exact months in a year, as their calendar had only 10 months, two months were missing. In view of this fact, one can easily imagine that the western world was not able to do exact dating in terms of days and month of historical events till 46 BC. That is why they don't have any accurate historical dating system in place before Christ and their entire history starts with Christ as the centre of history. Prior to Christ, their history is non-existent for want of a proper dating tool as a calendar. Moreover until the period of Gregorian reform on 24 Feb. 1582, their calendar was not perfect, as they could not develop a proper method to adjust the balance of 5 hours, 12 minutes and 48 seconds in a year as leap days or leap years. The reality is that it was not until 46 BC, but until 24 Feb. 1582 that historical datings done by westerners becomes arbitrary and contentious in absence of the adjustment of 6 hours' time and defining a leap year. On the other hand, the historical datings done by Indians since time immemorial in lunar months are exact. For example, Ram Navami (Chaitra Śukla Navami) is a perfect date handed over to us since millions of years, which cannot be disputed. Thus we see that India's Vedic calendar developed since time immemorial is an astronomical calendar which is quite relevant till date, whilst the western calendar underwent several changes and reforms till 1582 and yet remains arbitrary, so far as the days of months are concerned. The days like 30 or 31 assigned to various months have no scientific basis, but based upon the principle of 'might is right' which will become evident from ensuing lines.

One of the methods applied in formulating the synodic lunar calendar in the Vedic tradition is given here. If the period of 29.5 days of the synodic month is divided into 30 *tithis*, the quotient will be .983 which shows that the *tithi* is shorter by synodic day by .017 days (1-.983=.017) or .408 hours [.017 x 24=.408] or 24.48 minutes [.408 X 60=24.48]. Similarly, the length of a synodic month (29.5 days) comes shorter by solar month [365.25÷12=30.4375 days)] by .9375 days (30.4375-29.5=.9375) or 22.5 hours. On the other hand, we can say that with the passing of one solar month, a synodic month stands in excess of .9375 days or 22.5 hours and so with every 32.46 or say 32.5 solar months exceeds one synodic month. This excess or intercalated month was called *Adhika māsa* or *Mala māsa* during the Vedic age.

vedamāso dhṛtavrato dvādaśa prajāvataḥ
vedā ya upajāyate.

(*Ṛgveda* 1.25.8)

"That is the knower of the natural laws, knows the twelve months and the intercalary month."

The calendars formulated on this principle register dual months like two *Pauṣas*, two *Maghas*, etc. after 32.5 months and dropped the second intercalated month thereafter to allow the synodic months go in line with solar months. This was one of the methods applied in formulating the synodic lunar calendar in the Vedic tradition.

Another method was developed in view of the duration of *tithis* in lunar and solar years. This method used to be applied in the *Ṛgvedic* tradition as pointed out by *Ṛgjyotiṣa.* Accordingly, if the length of month and year are measured in *tithis*, the lunar year has a duration of 30 x 12= 360 *tithis* or 354 solar days (29.5 x 12=354). If we divide 360 tithis by 354 solar days (360÷354) we get that one solar day is equal to 1.0169 tithis. Thus a solar year will have (1.0169 x 365=) 371.1 *tithis* as compare to 360 of a lunar year. Thus a solar year has 371.1-360=11.1 *tithis* extra than a lunar year, which is called *Ṛtuśeṣa* in *Vedāṅga jyotisa* . On the other hand, we can fairly say that with the passing of one solar year, one lunar year stands in excess of 11.1 *tithis.* At the end of each solar year, this *Ṛtuśeṣa* of

11.1 *tithis* gets added and ultimately when the total of *tithis* exceeds 30, they constitute an additional month. Vedic people found out that the exact harmony of lunar and solar years takes place after 19 years where 7 such additional lunations take place. For instance, total *tithis* of solar year in 19 years will be 371.1 x 19=7050.9 and of the lunar year be 360x19= 6840. The excess *tithis* of lunar cycle in 19 years are 7056.6-6840=210.9 days. This number is exactly equal to seven additional months, i.e. 30 x 7=210 days. Perhaps these results inspired some schools of Vedas, as is obvious from Lagadh's *Vedaṅga Jyotiṣa*, to frame a *yuga* of 19 years with one leap year in each *yuga.* Lagadha also records 18 years with 371 lunar days and 19th year with 372 lunar days. Thus we see that some schools of Vedic thought observed a 19-year *yuga* cycle consisting of 235 synodic lunar months.

We also find the above method of harmonising the luni-solar calendar, in addition to 19, into the multiple of 19, such as 95 years in the *Yājuṣa* tradition of Vedas. But the 19 years luni-solar cycle remained a popular base.

Here it may also be noted that the time calculations based on lunar and solar cycles led to the origin of lunar dynasties and solar dynasties among the Vedic people in Bharat. Vedic civilisation in Bharat is the oldest civilisation on the Earth. The Vedic people from Bharat were the first people who inhabited the entire globe at various phases of times. These migrations from Bharat to various parts of Globe took place in the early ages as well as after nuclear and atomic war of Mahābhārata. High radiation generated due to nuclear war of Mahābhārata compelled the people from northern parts of Bharat to migrate to safer places. Thus wherever the Vedic people spread on the earth, they took along with them their Vedic heritage. Today also Vedic people on the various parts of Globe can be identified with the remnants of the Vedic heritage they are carrying along with them. For instance, the Vedic migrants from solar dynasties to Egypt are still known as the people belonging to Hamitic race and Vedic migrants from lunar dynasties are keeping their history of migration alive in the name of Semitic races in China, Arab, Babylon, Syria, Mongolia, Irak, Iran, Tartar, and Judea. The languages of Semitic people are known as Semitic ones and Hamitic people as Hamitic ones. The word Hema has corrupted from *sūrya* or *hema* as helios. The renowned city of the sun in Egypt is known as "Heliopolis" and the first king of Egypt "Menes" came from the race of Sun, Vaivasvata Manu. According to Pococke, the writer of *India in Greece* (P. 178) edited by the present author as '*Indian Origin of Greece and Ancient world*', first king of Egypt was associated to the Manu Vaivasvata of Patriarch of the solar race. Thus the entire Bharat was divided into lunar and solar dynasties and the entire world also got divided into Hamitic and Semitic races according to the solar or lunar class of the migratory people from Bharat. That is why the entire world is having the signatures of Vedic tradition carried either by the emigrants from lunar race or the solar race. Same is the case with the calendar systems prevalent in various countries.

"In China the descendants of King Pururavā", according to Col. Tod, *Annals of Rajasthan* (Vol. 1. p. 35) "The genealogists of China and Tartary declare themselves to be the descendants of 'Awar' son of Hindu King 'Pururawa', carried along with them the 19-year luni-solar cycle consisting of 235 lunations which represented the 19-year yuga of 235 synodic lunar months developed by Vedic people".

Yadus of the lunar race of Bharat now known as Jews migrated to Judea after the great war of India (According Pococke, *India in Greece*, p.22, 'The tribe of Yudha is, in fact, the very Yadu of which considerable notice has been taken in my previous remarks. According to Col. Tod, *Tod's Rajasthan*, p. 529, The annals of the Yadus of Jaisalmer state that long anterior to Vikrama, they held dominion from Ghazni to Samarkand). They also carried along with them the Vedic concept of luni-solar calendar based on a 19-year cycle with intercalations in 3, 6, 8, 11, 14, 17 and 19 years. The intercalation is made in the months of Adar (Feb.- March).

Vedic emigrants in Babilonia [According to 'Historical History of the World, vol.1, p. 89, 'The name of Babylonian storm God was *Matu* or *Martu* which, as we have seen, was the same as the Vedic *Marut* and must have been taken by the Paṇis and Cholas (of ancient Bharat) to Babylonia. Robert Brown says, 'This stellar and originally solar Ram stands at the head of the 10 antediluvian Babylonian kings...' See '*Origin of Week*' by Sham Shastri published in Annals of Bhandarkar Institute vol. iv,

Part 1, July 1922)] also used to bring a lunar year of 354 days into line with a solar year of approximately 365 days. This was accompanied by the use of intercalated month. The special name given to intercalated month was '*iti diring*'. But due to loss of memory and for want of connection with the mainstream in Bharat, the intercalation was operated haphazardly according to real or imagined needs and each Sumerian city intercalated months at its will. Later in 380 B.C., they succeeded in computing an almost perfect equivalence in a luni-solar cycle of 19 years and 235 months with intercalations in the years 3, 6, 8, 11, 14, 17 and 19 of the cycle.

The emigrants of the lunar dynasty in Arab (The name Arab is derived from Sanskrit '*arva*' meaning horse) followed the lunar cycle only for their calendar totally ignoring the solar cycle.

The Vedic migrants in Greek (Considerable evidence of emigration of Indians to Greece have been given by the author of *India in Greece*) also used a lunar calendar consistent with solar days. They completed the year in 354 days with 12 lunar months, alternatively consisting of 30 days and 29 days. But they lost the heritage and missed the seasons. King Solan, therefore, tried to introduce an intercalary month to harmonise the lunar year with seasons, but he was unsuccessful.

In Egypt, the Vedic emigrants (According to Col. Olcott, *Theosophist* for March 1881, p. 123, 'We have a right to more than suspect that India, eight thousand years ago, sent a colony of emigrants who carried their arts and high civilisation into what is now known as Egypt.') from solar dynasty used solar calendar. They kept their heritage intact and remembered that an year consists of 365 days. They divided the year in 12 months consisting of 30 days each and added 5 days at the end of the year. Yet the seasons rotated in a cycle of 1460 years due to a mistake of about 1/4 day in an year, because 365x4 = 1460 years. Modern astronomers call the period of 1460 years as a Sothic cycle.

The Parasoos, the people of Parasu Ram from Bharat who migrated to Persia (*India in Greece*, p. 45) also remembered the year of 365 days and so they also used to add 5 days after an interval of 360 days to constitute an year of 365 days.

Vedic emigrants in Rome (Both among Greeks and Romans) - the descendants of colonists from India, continued, specifically amongst the latter people down to and throughout the most historical periods (*India in Greece*, p. 142) totally forgot this division of zodiac in 12 parts. They used to count 10 months in a year and left some days uncounted. They begin with a fresh year with a full moon in the spring season as was in their memory. The names of their months were still driven from their original language, the Sanskrit such as unus from *eka* or *ūna*, duo from *dvau*, tria from *traya*, quinque from *pañca*, sex from *ṣaṭ*, septem from *sapta*, octo from *aṣṭau*, noven from *navan*, decem from *daśan*. The presently extant names like September, October, November and December clearly substantiate this fact. The term ambar is indicative of zodiac in the Sanskrit language. These 12 zodiac signs were known as 12 *ambars/rāsis* or stars and as a result, the names of months were coined by suffixing *ambar* /amber to the numbers. With the passage of time, first to six months were renamed either after some historical personalities or some specific historical events, though the last four months survived in their original form signalling their very origin to the Vedic system.

The Romulan calendar consisting of 10 months beginning from March as the first month up to December as last month remained in vogue till 452 B.C. when Numa Pompilius introduced the custom of inserting 23 days at the interval of two years. But the introduction of 23 days became the bone of contention between the religious heads. This was at last removed by Julius Caesar who with the help of Cleopatra, the empress of Egypt reformed the Romulan calendar after the fashion of Egyptian calendar. This reform was known as Julian reform or Julian arrangement of the calendar. The history goes like this.

Cleopatra in Altes Museum, Berlin

Cleopatra was the empress of Egypt. She was Greek by birth and belonged to the dynasty founded by Ptolemy, a commander of Alexander. Following the death of Alexander, Ptolemy declared his domain over Egypt. Cleopatra was from his dynasty. She was born in 3033 Kali era or 68 B.C. Following the death of her father, Ptolemy, the 11th, she took over the reign of Egypt along with her brother Ptolemy, the 12th. Nevertheless, she was overthrown by her brother in the war of succession. She formed an army in Cypress and tried to recapture power. In the meantime, defeated by Caesar and fled to Egypt, a brave soldier named Pompey was guillotined by Ptolemy. With the view to please Caesar, he (Ptolemy) presented his body to Caesar, but Caesar got annoyed at this. To make a good use of this event, Cleopatra hiding in the precious carpets purchased by Romans met Caesar and he was also taken in by her exquisite beauty. He captured the regime from Ptolemy and handed it over to Cleopatra. Afterwards, both Caesar and Cleopatra roamed about the banks of the river Nile for many weeks. Cleopatra gave birth to a child who was named as Caesarean, *i.e.* the Junior Caesar. When the queen of Egypt started domiciliating the palace of Rome as a queen-dowager, the artisans, economists and astrologers were sent for from Alexandria, the capital of Egypt. Roman taxation system was reformed and a new currency was introduced. Roman calendar was also amended by dividing the whole year into 365 days by adding two more months to the already existing 10 months, after the fashion of Egyptian calendar. Since the time of Julius Caesar

(46 B.C.), the Roman calendar used to have 12 months with the new year commencing with March as the first *amber* (month).

March continued to be the beginning or the legal year in England until the 18th century. In France, it was reckoned as the first month of the year until 1564, when by an edict of Charles IX, January was decreed to be the first month, so that the year may end with December, the month of Jesus Christ's birth. Scotland followed the example of France.

Julius Caesar

The history goes further and we are told that during this arrangement, the quintiles amber (fifth month of Romulan calendar) was named July by the Egyptian astronomers to felicitate Julius Caesar, as he was born on 12th of this month. This month was also assigned 31 days since Caesar was the most powerful king of Rome. On the other hand Caesar also, in turn, felicitated Cleopatra by constructing a temple of Goddess of Venus. Iluckily, Caesar died after two years in 44 B.C. and Cleopatra had to return to her kingdom, Egypt.

Julius Augustus

Later Julius Augustus also named the sixth *amber* (sixtilis) after his name as August and also assigned 31 days to his month, since he considered himself no less brave than Julius Caesar.

Mars, God of war and father of Romulus

Similarly, the first amber was named as Mars or March after the god of war. Actually, this was not only the beginning of the year but was the open spring season for waging war. That was why, it was named after Mars, the god of war. 'Mars" being the powerful god was also assigned 31 days.

The second *amber,* April, was named after Latin aperire 'to open' in allusion to its being the season when trees and flowers beginning to 'open' and is supported by comparison with the modern Greek word 'opening' for spring. It was given fourth place in Julian calendar.

The third *amber* was named Maius by Romulus in respect to the senators and nobles of his city, who were called Majores. Being associated with Majores, it was also given 31 days. It became the fifth in Julian calendar, which later came to be known as May.

The fourth *amber* was called Junius in honour of the youths of Rome (*i.e.* Juniares) who served

Romulus in war. Junius later came to be known as June.

Lord Ganesh

Januarius (January) and Februarius (February) were added at the end as 11th and 12th *amber* to the series by Numa Pompilius. Though in Julian arrangement their place was changed otherwise and they were considered to be the first and second months of name January is also identifiable with Sanskrit Ganesh. Ganesh of Indin an tradition is known as Janus in Roman tradition. Since Ganesh is the symbol of the beginning of every act in Indian tradition, *Janus* in Roman tradition *was also identified as* the god of beginnings and transitions, and thereby of gates, doors, doorways, passages and endings. He is usually depicted as having two faces since he looks to the future and to the past. The Romans named the month of January *(Januarius)* in his honour, which was also used by them as the first month of the year.

Statue of Janus in Vatican Museum

February was spelt by the Romans as Februarius. Surprising thing is that this also came from Sanskrit '*Pitara*'. Februarius is often derived from februum, which is nothing else but a corrupt from of Sanskrit word '*Pitaram*', meaning ancestors or forefathers or manes. It is to be noted here that Indians used to honour their ancestors and propitiate their dead during 15 days period ending *Pitṛyāna* (which is a period close to the vernal equinox or 21st March) as *Pitṛpakṣa* or *Śrāddha* or the period of manes. Although today the *Pitṛpakṣa* or *Śrāddha* is held close to the period autumnal equinox. This shift in the tradition has been elaborately explained in another treatise by the author titled as 'Scientific Significance of Indian Festivals'. In the same manner, februum, in Rome, is used as a thing for ritual purification. Most of the observances, in Rome, in this month concerned the dead or closure like that of India, reflecting the month's original position at the end of the year. The Parentalia was a nine-day festival honouring the ancestors and propitiating the dead, while the Terminalia was a set of rituals pertaining to boundary stones that were probably also felt to reinforce the boundary of the year. This is the actual history of the two months added to the Julian calendar.

In a practical sense, they still continue to be the last months. Since all the additions and subtractions are made to the last member of the series. Feb., being the last one, always adds a day in case of a leap year.

From the foregoing discussion, it can unhesitatingly be inferred that all the English months can be traced back to Sanskrit via Greek, Roman and Egyptian ones.

However, the above discussion sheds an ample good light on how the scientific arrangement of months of an year done by Vedic people got an unscientific and communal colouring at the hands of their immigrant in Europe and elsewhere. The assignment of 31 or 30 days too to various months of the year was done in tune with the slogans of 'Might is right' ignoring any justification or norms. Similarly the assignment of the status of last (12th) month to the 10th month simply because Jesus Christ was born in that month shows the communal bias of Christians to Christianise the whole world or allow the whole world to celebrate and practise the Christian traditions in disguised forms. Here it may not be out of context to inform everybody, even to our Christian brothers, that the word christ-mas do not signify the birthday of Christ, but the month of the birth of Christ. Since -mas is formed from Sanskrit term 'māsa' meaning 'month'. In Vedic times, the measurement of the month didn't owe either to rotation or revolution of the Earth but it owed to the Moon's revolution around the Earth. The following statement of Ṛgvedic *Ṛṣi* is noteworthy:

aruṇo māsakṛd vṛkaḥ (*RV.* 1.105.18)

Moon is the maker of months.

Yaska's commentary is noteworthy here.

aruṇaḥ ārocano māsakṛn māsānāṁ cārdha māsānāṁ ca kartā bhavati candramā vṛkaḥ

That is, Moon is the maker of months and fortnights.

In fact, the term *māsa* is formed from -mas of *Candra-mas* itself. *Māsa* means to measure - *māsā mānāt* (*Nirukta*, 4.27). Since the moon measures the constellations in the sky, the action of measuring is associated with Moon. Keeping in view the same measuring action, Greeks called it the Moon.

Here it would not be out of context to know as to how the lunar months were named in Indian tradition. Months in Indian tradition were formed on the basis of Moon's revolution around the Earth. Similarly, names of months were also given after the asterism which comes into conjunction with Moon at the end of *Pūrṇimā*, the full Moon day. For instance name of the month, *Caitra* has been given on the basis of a conjunction of Moon with *Citrā* constellation after the full Moon day. Here one more thing may be remembered that when Moon is in conjunction with a particular constellation at the end of *Pūrṇimā*, the Sun is housed 180^0 apart from the Moon or say the right opposite to Moon. Thus when the Sun is in *Aśvini* constellation, Moon will be in *Citrā* at the end of *Pūrṇimā* and the month's name will also be *Caitra* after *Citrā*. Thus one can see during the long span of time the calculation of year and its months came down to us in its most scientific form without any communal, regional, personal, racial or another type of bias. Let us come out of -isms and biases and be Vedic (scientific) in our practical life and in our attitude.

Made in the USA
Monee, IL
13 September 2026

62973760R00016